Enjoy Learning Homeopathy

An eightieth birthday celebration volume of homeopathic poems

www.EnjoyLearningHomeopathy.co.uk

Copyright © 2014 by J. M. English

Cover picture and illustrations for the Gelsemium poem, Mick Abbott.
All other illustrations, Cecil Holden

All rights reserved. This book or any portion thereof may not be reproduced or used in any manner whatsoever without the express written permission of the publisher except for the use of brief quotations in a book review or scholarly journal.

First Printing: 2014

ISBN 978-1-291-93115-0

Enjoy Learning Homeopathy
87 Greencroft Street
Salisbury
SP1 1JF
United Kingdom

www.EnjoyLearningHomeopathy.co.uk

Enjoy Learning Homeopathy

An eightieth birthday celebration volume
of homeopathic poems

Dr John English

2014

Dedication

To Farley Spink, who first introduced me to homeopathy.

Contents

Preface	ix
Acknowledgements	xiii
Introduction	xv

The Family White   3
   An Arsenicum Poem

The Rhyme of Bryan Albatross   17
   A Bryonia poem

The Concerns of Miss Cora Stickum   23
   A Causticum poem

Ode to Natrum Mary   31
   A Natrum muriaticum poem

Victor Nux's Nasty Day   37
   A Nux vomica poem

The Ballad of Jack and Jill   45
   A Pulsatilla Poem

The Confessions of Clara Cuttlefish   53
   A Sepia Poem

The Rhyme of Ivy Rusto   61
   A Rhus Toxicodendron Poem

The Collapsed Man   65
   A Carbo-Veg Poem

No Joy for Jasmine!   71
   A Gelsemium Poem

Preface

A few years ago my father faced a serious, high-risk operation. Seeing him for a few days beforehand, I was concerned by his uncharacteristically low spirits. Usually, he is remarkably calm and philosophical about life's Great Matters. But something was bothering him. Catching a quiet moment, I asked him what it was.

'It's my poems,' he said. 'If I go, they go too.'
'Well,' I replied. 'I've been asking you for years to let me help you. I know you want to finish them, but surely there's a case for sharing what you already have?'

What I didn't realise, with my innocent offer, was the sheer quantity of material involved. Over the forty-odd years of his teaching career, my father has stashed away homeopathic teaching materials in boxes, dog-eared cardboard folders, more recently on his computer – some are still on the backs of envelopes. Setting up a website and bringing the materials to light has been the homeopathic equivalent of cataloguing Aladdin's cave. I discovered at least five different sorts of quizzes covering all the major remedies (perhaps a hundred quizzes in total), vastly detailed mind maps in various stages of composition for more than thirty remedies, a number of articles and commentaries, and above all, over forty poems and prose creations covering thirty or so different remedies: all this, dedicated to the fine art of learning homeopathy.

Of course, a prophet in his own land is rarely recognised. As children, we knew about my father's different interests, but mostly dismissed them as, 'another one of dad's things'. Words like 'macrobiotic' or 'psychosynthesis' were more familiar to us than 'antibiotics' or 'vaccine'. Although we were all fully vaccinated (and did receive conventional medicines if we really needed them), if we were ill, we were usually treated with homeopathic remedies, kinesiology or acupuncture. It was only later that we discovered allopathic medicine was the norm.

We knew dad offered homeopathy to his patients without realising just how unusual it was for a regular doctor, a General Practitioner in the National Health

Service, to embrace so many therapies and techniques. More significantly still, he made them readily available at his GP surgeries. The authorities noticed something was happening, asking questions about his annual drugs bill – why it was so low! In 1981 my father was awarded the prestigious Fellowship of the Royal College of GPs, in recognition of his outstanding achievements.

In a general practice set amidst suburban housing estates in the 1960s, 70s, 80s and early 90s, this approach was radical. His research (for example on the effect of relaxation meditations on pregnant mothers) was way ahead of his time. Recently, mindfulness has been endorsed by the NHS for certain specific conditions; my father was teaching Autogenic Training (a similar practice) – and proving its efficacy – almost half a century ago.

My father's dedication to his patients permeated our home. Evidence appeared in the shape of small gifts; chocolates, wine or a strangely wonky vase. We owed the magical arrival of our cat, Smokey, to a patient who couldn't care for her. We laughed at his quirky tales, such as the recommendation to a poorly patient to 'eat kippers with marmalade for breakfast' (whether this worked as a cure, history doesn't relate!).

Surrounded by his busy General Practice, not to mention his busy family, with five tumbling children and a dog, my father needed a creative space of his own. He grabbed spare moments. I remember him sitting bundled up in his huge coat and furry hat on the piano stool practising a Chopin waltz before we went out, while mum raced around getting us all ready. He also wrote verse. No birthday card was complete without a special rhyme just for us ('The time is ripe when father oughta, eulogise his elder daughter ...'). Honorific doggerel still graces Grand Occasions, as close family friend, Peter Collings-Wells (a keen caver), discovered on his 50th birthday:

> *There cannot be many, when planning a rave*
> *Would choose for the venue a cold, damp cave!*
> *Whatever the plan is, we're sure, if you're nifty*
> *You'll find a good way to enjoy being fifty.*
> *And we nearly forgot – Good heavens above*
> *To send with this card a great parcel of love!*

Even humdrum activities might provoke the muse. Here, dog-sitting for his niece's Westie, and constantly wrong-footed by the scuttling hound, my dad ruefully remembers the 'honest idiocy of flight' of The Butterfly – in this neat re-working of Graves' poem, Flying Crooked:

The Ollie Dog
(with apologies to Robert Graves)

*The Ollie-dog, as white as chalk
Whose honest idiocy of walk
Will never, now, it is too late,
Master the art of walking straight.
He has – who knows so well as I?
A just sense of what not to try.
He lurches here and here by guess
And cat and scent and wilfulness,
And never does he lose his grip,
On knowing how to make us trip!*

For as long as I can remember, my father has burst into verse or into song when something reminds him of a line. Even drinking tea can be accompanied by an English madrigal ('and poured it through a strainer, poured it through a strainer ...'[1]). And just as his songs and poems are part of our lives, so parts of us enter his poems. Reciting his Silica drama for his students, my father dons the long woolly scarf knitted years ago by my sister, to illustrate the dread of cold.[2] The ginger cat which creeps into his Gelsemium poem (p. 71) comes from my grandfather. Arriving one day at the hospital to visit our sick Grandma, Grandpa placed a loveable soft toy cat in her arms.

---

[1] The madrigal by Giles Farnaby (c. 1563 –1640) is one we often sang as a family: *'Simpkin said that Sis was fair, and that he meant to woo her. He set her on his ambling mare, all this he did to prove her. When they came home, Sis flotted cream, and poured it through a strainer, but said that Simpkin should have none, because he did disdain her.'*

[2] Sadly too long for this volume, the Silica dramatisation appears on the website: www.EnjoyLearningHomeopathy.co.uk

'Who is this?' Grandpa crooned. 'Soonby!' he declared, answering his own question. 'Because you'll soon-be better ...'.

Over time, I've come to value the creativity and passion of my father's verses. The very basic rhythms and rhymes in the homeopathic poems might mislead you into thinking this is simple versifying. In fact, it's the result of extensive research into learning methods (once again, distinctive for the time, and ahead of the trend), immense attention to homeopathic detail, and deep consideration of the remedies he was wanting to portray. A flavour of this emerges in the following Introduction, where my father describes the ingredients necessary for a successful homeopathic poem. Even for a non-homeopath like myself, the poems evoke a strong atmosphere that lifts them beyond mere mnemonic devices to something more. A growing number, read by my father with animated illustrations, are now available on YouTube[3] – so you can see for yourself.

My father continues to create new materials for his website. It's a true legacy, offered freely through his personal generosity. But I know that, deep in his heart, he has always wanted to hold a book of his poems in his hands. Now, for his 80th birthday his five children, cousins and close family friends are proud to offer him this celebration volume. It's a small selection of his work, maybe; but a contribution to homeopathic literature nevertheless. To a remarkable father, uncle and friend, whom we love dearly,
we wish you a happy 80th birthday!

Elizabeth English, 2014
With Peter, Helen, Jeremy and Quentin

This book is a gift from all the family: Peter, Doro, Arkady, Sebastian and Tirion; Helen, Bill and Laura; Elizabeth; Jeremy, Joanne, Lily and Leo; Quentin, Joanna, Chloé and Soren; from cousins Alison, Alex, Edward, Jamie, William and Nikki; Andrew, Ellie, Rose and Ariana; and from dear friends and almost-family, Peter Collings-Wells; Caroline, James, Mollie and Harry Gardner; and John Muir and family.

---

[3] www.youtube.com/user/EnjoyHomeopathy

Acknowledgements

Above all, gratitude to my mother, Wendy English, for her love and care of her poetic doctor husband, for almost 60 years now. Her cousin, Cecil Holden, although he passed away many yeas ago, still holds a place in my father's heart for his witty illustration of many poems. In the early days of the website project, Peter Collings-Wells gave long hours of his professional photographic expertise to prepare the cartoons for use; as well as his warm belief in the value of dad's work. This volume would not exist without the tremendous support, creativity and resourcefulness of Shani D'Cruze, who now coordinates the Enjoy Learning Homeopathy project, as well as continuing to edit the material and devise all manner of technical wizardry to bring the mind maps to life online. Laurie Girling pioneered the videos with huge doses of time and talent, adding a whole new dimension to the poems. With extraordinary generosity and dedication, artist Mick Abbott has taken up the challenge illustrating other poems, thoroughly immersing himself in the character of the remedy – which comes across in the strength and beauty of his paintings. We value the contribution of Alison Shakespeare, mostly on the quizzes, the designs by Ivo Mesaros, and the expert guidance of Matt Heselden as he guides us into the world of social media, helping to make these materials better known.

Introduction

Why write a poem?
By Dr John English

Over the many years that I taught homeopathy, I devised all kinds of learning materials to help my students learn. When creating these teaching aids, I would sink as far into the remedy in question as I could, thinking about it, reading different textbooks, noticing if and when it applied to me. This kind of active learning helps the material to stick, and become part of you, as it needs to do in homeopathy. You read a very good Materia Medica textbook, working late into the night. You meet all these colourful symptoms: let's take, 'tulips growing from the umbilicus'. Next day you think back on last evening's work and that interesting symptom – and for the life of you, you can't remember which remedy it refers to. Oh dear!

I've met so many students who nod enthusiastically when I tell them this. It's my own experience too. To help myself learn, I studied Suggestopaedia. In my early days as a tutor at the (then) Royal London Homeopathic Hospital, I tried to absorb its lessons. Suggestopaedia suggests that rhythm, rhyme and humour aid learning. This was not at all the direction serious modern poetry was taking! But my purpose was to learn, and to help my students learn. I wanted to make learning as enjoyable, as fun and exciting, as possible.

I turned to poems and pictures because we have visual and auditory memories. We respond to rhyme and rhythm, fun and humour. We learn better when we relax and engage our creativity. A good homeopathic poem, according to Suggestopaedic principles, should contain as many strong rubrics of the remedy as possible. That's why my Arsenicum poem, The White Family Album – the longest poem in this collection – has over a hundred rubrics hidden within its verses. If you want to see the repertorization for these poems, you'll find it on my website.[1] A memorable poem must rhyme and have an easy rhythm. Sometimes a nursery rhyme suggests itself. For example Pulsatilla's

---

[1] Go to www.EnjoyLearningHomeopathy.co.uk/start-learning/remedy-poems-pictures

Jack and Jill poem and Gelsemium's No Joy for Jasmine use nursery rhyme format. A catchy title is essential.

There is also a place for memorable, dramatic lines and descriptions which stretch our understanding of the remedies to the limit, as in my Sepia poem, The Confessions of Clara Cuttlefish. Some of my poems aim to catch the underlying character of the remedy, as in Victor Nux's Nasty Day, for Nux vomica. Others economise on words, creating a picture with the fewest possible lines. A good example of this is The Rhyme of Ivy Rusto (for Rus toxicodendron). Where the Materia Medica indicates strong feelings, a poem can sometimes capture them more completely than prose, as I found in my Ode to Natrum Mary, for Natrum muriaticum. On occasion, as I absorbed myself in the remedy, the poems would almost write themselves.

Of course, all this is enhanced by visual learning. In my case, since I can write poems but not draw pictures, the poem came first and the cartoons followed. Good pictures reflect the remedy. They need bright colours, somehow just the 'right' colours for the remedy in question; and I always chose a visual symbol based on the remedy to repeat throughout the illustrations. I was very lucky to have such a good cartoonist, Cecil Holden, a relative of my wife's. Unfortunately he passed away many years before this volume was printed. The most recent illustrations for my poems are by Mick Abbott, an accomplished artist with a very different but equally effective style. For my website, the poems and illustrations have been given an added level of memorability and fun through Laurie Girling's animations.

I have found that writing a poem for a remedy makes that remedy more real for me. The writing itself is a learning tool! So while I hope you enjoy these poems, and enjoy learning the remedies through the poems, I would encourage you to follow my example and bring your own unique creativity to the complicated task of learning homeopathy. You will soon discover for yourself what helps you learn homeopathy. And if you can share that with others, so much the better!

Dr John English, 2014

www.EnjoyLearningHomeopathy.co.uk

The Poems

The Family White
An Arsenicum Poem

I.

Come all and listen while I tell
A tale you should all know full well,
Of Arthur and of Amy White,
And of their family's sad plight.

The cause of it, do not forget,
Was living in the cold and wet.
Their diet, too, was never good,
They ate more soft fruit than they should,
Compounded with tobacco smoke.
No wonder they would wheeze and choke!

They knew 'twas wrong, and therefore built
A life of anxious fear and guilt
In which, despite their every care
Was hopelessness and black despair.

The Family White

But first of all I need to say
They are not always seen this way;
In time we will portray the worst,
The better features must come first.

Their house and garden might be seen
Pictured in a magazine.
Furnished with so great a care
It doesn't have a 'lived in' air.
A place where every flower grows
In formal, neat and tidy rows,

So horrified at any dirt,
No spot of it on any shirt.
As if such cleaning could begin
To wash away their inner sin.

II.

Arthur's keen and lively mind
Is of the intellectual kind.
His logic leaves unturned no stone,
He picks each problem to the bone.

Perfection is his worthy aim,
And if not reached, then he will blame
Himself severely, and he'll rail
At anyone who made it fail.

Immaculate in every way,
His dress, surroundings, thoughts, array
Themselves so tidily they give
The feeling they're too good to live.

From somewhere deep within, a drive
Forces Arthur still to strive
When other peoples' work is done,
If his life's battle's to be won.

Restlessly it makes him pace
The floor, as if all life's a race,
Nor mind nor body ever still
While there's a spark of hungry will.
Then at the weekend he may stop,
And with a fearful headache drop.

The Family White

His staff are seldom heard to moan
At his authoritative tone,
Though critical, he's always fair
If work's well done, so they don't care
To flout the least of his commands
And have to face much worse demands.

His business sense is very good,
His projects flourish as they should.
From humble start, as mere inspector,
Success as Managing Director
Demonstrates achieved ambition –
It's a family tradition!

Though anxious now about his health
He shows some signs of increased wealth.
His (also Ars.) finance advisor
Eggs him on to be a miser,
Also to arrogance and pride.

If with challenge now you tried,
Or criticism of any kind
You'd see a different sort of mind,
For these he cannot take at all,
Or anything that makes him small!

III.

Amy's ambition now appears
Focussed on family's careers.
A disciplinarian at home,
Yet in their interest she'll roam
The country till she's tried
And found the best. Then, satisfied
She takes the children off to schools
Whose spotless uniforms and rules
Accord most with her own desires.

To more than school, though, she aspires,
And for their lives to be complete
Her children now must all compete:
Piano, violin and dance,
Brownies and Scouts. If there's a chance
To win some special scholarship,
No effort spared, she'll make the trip,
And hours spend in her smart new car,
Chauffeuring children near and far.

At home, while she may nag and scold,
Which neighbouring Mum would make so bold
To criticise her darling brood?

The Family White

She'd boastful turn, or even rude
In their defence. Put to the test
Her children have to be the best.

Yet, strangest thing I have to tell:
She never thinks they're really well.
The herald sniffle of coryza
Alertly heard, she deems it wiser
To pack the poor kid off to bed,
Lest in the morning he'll be dead!

"Pneumonia at least!" she'll say,
"Please doctor, quickly! Come this way!"
Her 'please' is an imperious 'must':
Her doctor has to earn his crust!

An Arsenicum Poem

IV.

It's midnight round to one or two
That little Willie White is due
To get whatever his next plight is,
Asthma or gastroenteritis.
Whichever form his illness takes
With striking suddenness he wakes,

And then, with swift acceleration
From fearful, anxious agitation,
Too gross, 'twould seem, should one observe
Than his apparent ills deserve,
Restless to his mother turns,
And lies exhausted on her bed,
Wishing and fearing he'll be dead!
Upon his pale face a sweat,
Then cold and shivery he'll get,
Relieved when mother makes him warm
– If he's running true to form.

The Family White

Now, should you not be a believer
In the Arsenic Album fever,
I'll tell you more about young Willie:
He starts by being extra chilly,
Till violent rigors rack his frame,
And icy wave sensations tame
His spirit. Then his doting mother
With heated drink helps him recover.

But not for long! Soon burning heat
Will make his misery complete.
"Help, help!" he cries, his mind in turmoil,
"My blood is just about to boil!
There's bees and wasps around my bed!
Why are they buzzing round my head?
And over there that horrid shape
Must be a thief! Help me escape!"

Suddenly he's through the door
To another bed on another floor;
Then, breaking into copious sweat
Cold and exhausted soon he'll get,
Which leads him to a raging thirst
For icy drinks, till he would burst.

But do not, please, form the belief
That drinking will give him relief.
That which can quickly downward plummet
Just as fast returns – as vomit.
Thus does he change, from spell to spell,
Till he'll eventually get well.

V.

Now, as if looking through a glass
I'll tell of Arthur's Auntie Floss.
She lies, knees drawn up, on her bed,
Believing that she'll soon be dead,

Which really will be no surprise:
Her sallow, pale, sunken eyes
And wasted, shrivelled dried-up skin
Bespeak the plight that she is in.

The flesh has gone that once did grace
A full and quite attractive face,
Her bony knee and matchstick arm,
Protruding ribs and joints alarm
Her family and doctor too,
Who know the pain that she's been through.

"If only she would eat," they say,
"Or she will surely waste away.
The tastiest morsels she could choose,"
But in the end she will refuse.

She feels so sick with just the sight
Or smell of them, although she might
Just sip a little drink of water,
Proffered by her anxious daughter.

The Family White

The wracking bouts of pain, which burns,
And comes in periodic turns
Is at the zenith of its power
At midnight's dark and dreary hour.

Attacks her then like red-hot pins –
A punishment for bygone sins?
Of strangest features this is chief:
More heat will bring her pain relief.

It isn't only auntie's gloom
That spreads itself throughout the room.
The pungent, cadaveric smell
Which we, the doctors, know too well,
Is more than just the diarrhoea
That can incontinently appear –
Reminds us all of Adam's curse:
The diagnosis can't be worse.

Her body is too weak to try
To move, yet restless, anxious eye
Beseeches everyone to stay.
She's frightened when we go away.

VI.

Treating patients of this kind
Lingers in the doctor's mind.
In anxious, though commanding tone
They question you upon the phone.
What do you do? Where qualified?

Eventually, when satisfied,
So you will understand them better,
Their history, neatly in a letter
Precedes the consultation date.
Early there, they sit and wait
Inspecting all with eagle eye,
The room, staff, patients, all espy
Fidgeting, till from smart brief-case
Extracting work with which to grace
The lagging time till consultation.
What a trying confrontation!

He tells the story his own way,
Not allowing you to say
More than "yes", "indeed," "ah so!"
Through endless detail he will go,
Complete with full interpretation

The Family White

Told with conviction and elation.
Then, when all has been confided,
He'll tell you next what he's decided,
How to investigate, refer –
And only then can you confer!

His anxious fear is plain to see
His speech and body both agree,
Gaze gimlet-piercing through and through,
Getting ever nearer you,
Leaning forward on his chair
His hands are clasped tight in despair.

As the meeting nears its end
More anxious energy he'll spend
To keep it going. Now he'll bring,
Repeatedly, "Just one more thing!"

And afterwards, to make quite sure,
He'll button-hole you at the door,
Giving not the slightest heed
To other patients, and their need.

An Arsenicum Poem

Though the pathology he hates,
Just have a care when it abates,
He'll view your therapy's achievement
As if it were a huge bereavement!
He so loves talking of complaints
He'd try the patience of the saints.

Accompany what you prescribe
With reassuring diatribe.
With ills so bad he burns in hell,
And he demands you make him well.
You! You're responsible for his plight
So just make sure you put him right,
Or else! And in the end th'unspoken threat
Will make you anxious, make you sweat!

The Rhyme of Bryan Albatross
A Bryonia Poem

Bryan Albatross, best known as 'Bry',
Kept a shop, he was in DIY.
When business was bad he was sullen, or sad,
Bad-tempered and worried. He'd try

To gather his thoughts in a plan,
As befits a booming businessman,
Making sense of his plight, but try as he might,
His figures would never quite scan.

His head is stuffed full of ideas.
To carry them out would take years,
But has he quite got it? He's bewildered, besotted,
Which justifies all of his fears.

The Rhyme of Bryan Albatross

Though sober his dress, and quite proper,
To impress each potential new shopper,
His sour greasy sweat and his dank hair will yet
Cause him to come a great cropper.

In fear for his future quite lost,
He'd endlessly dwell on each cost,
Till, energy used, and brain all confused,
He went for a walk in the frost.

When there's high pressure cold and it freezes,
Bryan Albatross comes out in sneezes:
No need to be told, it's the start of a cold,
And a series of much worse diseases.

Their onset is apt to be slow:
Dry nose, nothing comes with a blow,
Or water, or blood may come in a flood,
As the symptoms continue to grow.

He complains that he feels burning hot,
Tho' to touch, if you dare, he is not,
His bluish red flush is less like a blush
Than the sweat of a simmering pot.

A Bryonia Poem

Or with dry burning fever he's cursed,
From nine until midnight it's worst.
Sometimes a chill will make him more ill,
And his head is as if it would burst.

In delirium next he is caught.
He mutters and chunters, he ought
To respond, if you ask, but he snaps, so the task
Is one which is better not sought.

For Bryonia this is the test:
The troubles descend to the chest,
Soon setting off a dry tickly cough
And a hoarse voice that's better for rest.

He's a sharp cutting pain on the right
Of his chest, that comes on in the night.
He lies on that side, for he cannot abide
Any movement. He holds himself tight.

He has to, for this is his thesis:
That his chest will soon fly into pieces!
So, thorax or head he holds still in their stead,
As the pain from his problems increases.

The Rhyme of Bryan Albatross

Be it sharp, shooting, throb, stitch or burst,
He usually has a great thirst
For water with ice, gulped down in a trice,
Through lips swollen, cracked, dry and pursed.

Pericardium, pleura or joint,
Synovial membrane's the point.
Appendicitis, then peritonitis –
Whatever his fate might appoint.

Bursitic joints, swollen and red,
Become gout or arthritis instead.
Relieved, beyond measure, by stillness, warmth, pressure;
Sciatica by lying on it in bed.

A Bryonia Poem

You haven't yet quite heard it all:
Poor Bryan's in trouble with his gall,
And he's heard to moan in his stomach's a stone,
And his stool is too large, or too small.

There once was a Bryan Albatross,
A rolling stone gathers no moss.
He's happy at home, never wanting to roam.
His history: it's profit and loss.

The Concerns of Miss Cora Stickum

A Causticum Poem

It isn't right, it isn't right!
The world is an unhappy place.
Restless I lie awake each night
Pondering each new disgrace.

Outrage well beyond belief
Each day's reported on TV,
There's so much strife and so much grief,
I feel each hurt as if it's me.

With friends and relatives I'm heedful,
And, when they are old or ill,
I'll do for them whatever's needful
And risk giving myself a chill.

The Concerns of Miss Cora Stickum

I cringe and weep at each injustice,
Sympathise with each victim dread,
The thought, when all has been discussed, is:
The perpetrators should be dead!

Though timid and anxious now, the aspect
This thin, sallow body of mine conceals,
It allows me to do my work and protect
All of my tender, creative ideals.

My viewpoint is very firm and dogmatic,
Supporting every underdog cause.
People have sometimes called me anarchic -
Because I don't give a damn for their laws!

A Causticum Poem

I'll take up politics and join
The Social Conscience Party's call,
To this I'll give both time and coin,
To this I will devote my all.

I couldn't be a mere spectator
I had to join in, and belong;
Soon they called me a dictator
For chivvying them all along.

I'll organise anti-government petitions
And join in every protest march,
Don't care if they call it sedition,
Their attitudes are stiff as starch.

The Concerns of Miss Cora Stickum

The first time I went it was raining.
Warm and dry in my coat, I felt good.
The next, dry cold wind had me straining
My energy more than it should.

All day in icy wind I shouted
Banged a tin can with my hammer,
But when it was my turn for speeches
I could only stand and stammer.

Committed then I had no choice,
The day's events must take their course,
They took as well toll of my voice
And left me feeling very hoarse.

The mirror next day caught me snooping:
Was that really me? Oh, hell!
My face, right side, was flopped and drooping:
Paralysis, named after Bell.

Vindicated thus was my foreboding;
Justified, that I should darkness fear,
But reckless that my system was o'erloading
By doing that which to my heart was dear.

This ill's a chance! I can indulge desires:
Smoked meat and salt, cold beer to slake my thirst;
No sweets, thanks. Yuck! Some food I fear conspires
To make me ill: bread, fat, meat, and coffee is the worst.

Since I've found my true vocation
Answering the Lord God's call.
I'm filled with spiritual elation,
Fight the good fight, and walk tall.

Years later, now I'm getting older
My ills have grown more than of yore,
I'm timid, anxious now, not bolder;
More tears, more peevish, nerves more raw.

The Concerns of Miss Cora Stickum

The theme of my life has been trauma,
The mental sort, life full of grief;
I've been a reluctant performer
In support of my ardent belief.

It's worn out my quota of zeal,
So tired out now, and so weak;
Quite hopeless and low, and I feel
As paralysed now as my cheek.

I'm dizzy, and stagger on turning,
Must wake up to blow my blocked nose,
Pains everywhere, tearing and burning -
I'm not daunted at all by these blows!

My eyelids now droop and won't open,
Must stand when I go to the loo;
I tremble as much as an aspen,
My bladder is troublesome, too.

A Causticum Poem

When youthful I had enuresis,
Wet beds every night, to my shame,
But now it is more a paresis,
A cough, or a laugh is to blame.

My limbs are all contracted, stiff,
I feel each draught, am icy cold,
Can't think straight, each little tiff
Leaves me prostrate. I'm getting old.

Sad thoughts every day come o'er me.
Why do I live? My mind's on that,
And all those dears who've gone before me:
Ma, Pa, comrades – e'en the cat

Ode to Natrum Mary

A Natrum muriaticum poem

Natrum Mary, quite contrary
How does your garden grow?
Each cockleshell a private hell
And little tasks all in a row?

Self-effacing care you're placing,
Pouring out sympathy;
Hiding your moans and inner groans.
From conscience never free.

You're not too old to be consoled.
Why don't you have a cuddle?
Your tears are shed at night in bed –
Your feelings? In a muddle.

Ode to Natrum Mury

Nature will bring sweet flowers in spring,
Which fill the heart with gladness.
What others enjoy in you will cloy
Into a bittersweet sadness.

Music and art should play their part,
Good company beguile.
Poor lonely one, instead of fun,
False laugh and forced wry smile.

What could inspire such awesome ire?
So quarrelsome, such malice.
Sometimes you're quite consumed with spite,
A witch's burning chalice.

Your spotty face is no disgrace;
Your neck can't help being scrawny;
Your greasy hair, (despite your care)'s
A pleasant shade of tawny.

Cheiropomphylyx will itch
Like little grains of sago.
Fingertips white, that sometimes might
Be typical of Raynaud.

A Natrum muriaticum Poem

A sudden cold, now we are told,
Has brought you to your knees.
'Twas something sad that made you bad,
Eyes sore, and sneeze and sneeze.

It comes with chill and feeling ill
Before the clock strikes noon.
Tho' you perspire before the fire
You'll not get better soon.

Your poor red nose, which frequent blows
Makes sore, from salty water.
Catarrh like white of egg is quite
Profuse, and gives no quarter.

Herpes simplex upon your lips,
Which dry with central crack;
You might well burst from unquenched thirst,
It has you on the rack.

You're full of woes from head to toes:
One such is constipation.
It's such a pain to have to strain
With rectal prolapsation.

Ode to Natrum Mury

The woman's curse could scarce be worse,
For you it was thus named;
In blackest mood the wrongs you brood on
Make you feel ashamed.

To talk of sex will doubtless vex
You, cause a scarlet blush.
Sex to avoid, excuse employed,
Such as: "I've got the thrush!"

You wonder why with evil eye
Relationships are haunted.
With grave self-doubt you're put about,
By inner critic taunted.

He'll set in train a fierce migraine
With his judgemental clamours.
The pains which strike are sometimes like
A thousand little hammers.

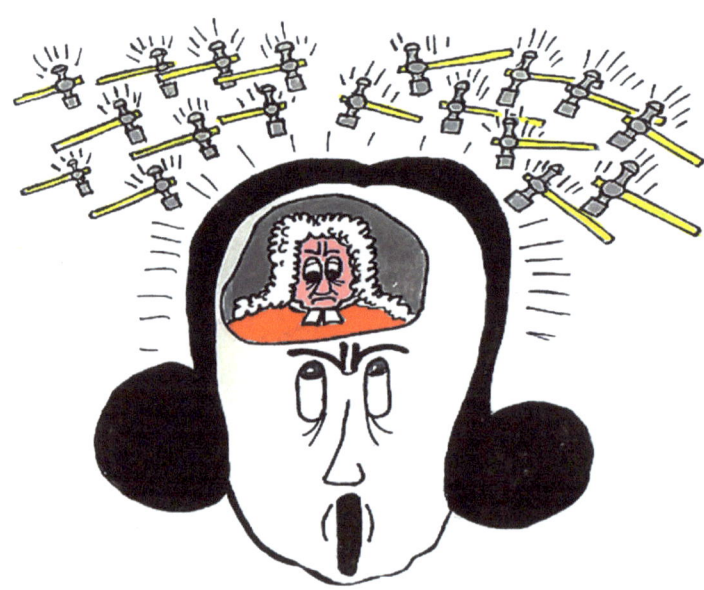

A Natrum muriaticum Poem

Trouble sown before you've grown
To full maturity.
It's shadow cast, and binds you fast –
Bizarre security.

Childhood loss and hurt, a cross
To bear has been your lot.
Hurt to repel, you built a shell
To live in, or to rot.

Your anxious fears and ready tears
Represent you badly.
You do much good – more than you should –
So why behave so sadly?

Complaints from grief: these are the chief
Amongst your list of ills.
To make you wise, may I advise
And save you further spills?

For looking back upon her track
Lot's wife was turned to salt.
So now you know, just forward go
And you'll avoid this fault.

Natrum Mary, quite contrary,
How does your garden grow?
With wishing wells and wedding bells
And salt pillars all in a row?

Victor Nux's Nasty Day

A Nux vomica poem

I'm in a hurry, mind your backs!
Put those cases on the racks!
As usual, the train is late,
I shall never make my date.

Here at last, the office door,
Sweep those petals off the floor!
Where's my secretary? Gaye!
What is there to do today?

Appointment lists? We start at eight?
Never finish at this rate!
At ten there are three items listed,
Why is that? Oh, I insisted!
I only have myself to blame
For booking Mr. What's-his-name.

Victor Nux's Nasty Day

Busy till there is at one
A working lunch with Mr. Lun.
Which restaurant? Oh, I'm in luck
I'll order special Peking Duck.
I 'spect I'll eat more than ought'er!
The thought of it makes my mouth water.

Ah! the coffee. Put it there.
Not so distant! Near my chair.
This place is an untidy mess
You should know this causes stress.
Put my papers in neat piles
Then greet the visitors with smiles.
While you're at it, clean those shoes!
There'll be some row if you refuse!

I've five minutes for this report,
Should be much longer, life is short.
Shut that door, and please keep quiet!
That noise sounds like some sort of riot.
God! it fills me up with anger,
Ears assaulted by that clamour!
Can't think straight with all that noise
Do I employ a mob of boys?

A Nux vomica Poem

The more there is upon my plate
The harder 'tis to concentrate,
And then I start to get the shakes
And make incredible mistakes.
It's still too early for a drink,
A whisky now would help me think.

Stomach pills, for goodness' sake!
That meal has left me stomach-ache.
Ever since the luncheon ended
I've been feeling quite distended.
And later, you may hear me groan
From pain as if there were a stone.

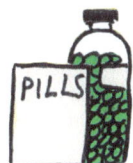

I never learn, and always rue,
Quickly, I must find the loo!
Such colic, as I sit and strain,
So little out, more must remain!
To make it worse I'm getting drowsy
It's hard to work when feeling lousy.

I say, that office girl's a corker!
I must see if I can talk her
Into something. Make a date;
What a thought! I'm getting sexy!
...Saves me getting apoplexy.

Victor Nux's Nasty Day

What's on tonight? A dinner date:
The Englishes? Their food is great,
And they are generous with the booze
What better could a person choose?

Success all round, this is the life,
To crown it, I've a gorgeous wife.
Some people think I give her hell
But there's another side as well,
I'm at my best into the night,
Our sex life makes it all alright.

Thanks a lot! A splendid dinner!
Roast beef surely is a winner.
Glad you served it with the fat,
I've quite a taste for some of that.
Brrrr! It's cold outside the door,
A summer coat, wish I had more.
An early frost, wind from the east,
Unseasonal, to say the least.
One for the road, I'll have instead,
Another on my way to bed.

A Nux vomica Poem

Oh God! Turn down that bloody light!
Have some sense, it's much too bright!
Its what? The same as every day?
It can't be, it's just what you say!

Oh hell, I'm feeling really ill
A hangover, and p'raps a chill
I feel it up and down my back,
A draught is coming through that crack
Which penetrates my very bones,
Adds to my ills, adds to my moans,
Makes me ache and shake and shiver
While last night's drink will rot my liver.

Quick, just go and get a pail,
A pain as if there were a nail
Hammered hard into my head
While straining to be sick in bed.
Too late to wish that I'd stayed sober,
Now my brain keeps turning over
Pain from my occiput is tearing
Muscles in my neck, and baring
All my sensitive nerve ends.
Touch me not! I'll make amends
Truly, it'd take a lot'll
Make me go back on the bottle.

Victor Nux's Nasty Day

Tchooo! a body-racking sneeze
Has come to make my ills increase.
Nose, a creeping, crawling tickle,
Now it's morning starts to trickle,
Soon to be a wat'ry flood
Mingled probably with blood.

Now my throat is getting rough
And sore, as if that weren't enough
Hacking cough, I'm going west,
Something torn loose in my chest.

Now another pain, infernal
Soreness that is retrosternal.
From my arms and legs I shiver
As I'm gripped by rising fever.

These blankets ought to keep me warm,
Wrapped right round my aching form
But fail to achieve the task.
For deliverance I would ask
From the chains of fever's cage
Petty things which make me rage:
Every little draught of air,
Every tick the clock makes there,
Eyes that cannot stand the light,
Even smells which would delight,
Normally, now cause me ire
And from which I will perspire.

A Nux vomica Poem

Suddenly, beneath this sheet
I am intolerant of the heat,
Yet, if I poke out one toe
As if I'm naked in the snow!

Who's there? You must be Dr Jones.
This cold eats right into my bones.
Quick, find some balm to ease this chill,
I've work to do, I can't be ill!
Is there no way to cure this flux?
What's that? "A tiny dose of Nux?"
Do you know, I think you're right,
In a minute I'm alright!
Really, you must stay and dine,
Please join me in a glass of wine!

The Ballad of Jack and Jill

A Pulsatilla Poem

Blue-eyed Jill went up the hill,
The miller's pretty daughter.
She took a pack, and went with Jack,
To fetch a pail of water.

Hand in hand they went as planned,
From stuffy indoors free.
The open air erased all care,
They would enjoy their spree.

They went at dawn on a sunny morn,
Their hearts were full of laughter,
For on the way they'd stop and play –
And fill the bucket after.

The Ballad of Jack and Jill

Lots of talk, and uphill walk
They soon found very tiring,
So Jack and Jill sat on the hill,
Both hot, but not perspiring.

All around upon the ground
They saw a purple flower,
Which bowed to them on slender stem
A welcome to their bower.

A gentle breeze soon rocked the trees
And flattened Pulsatilla.
A sudden chill affected Jill,
The daughter of the miller.

She soon had lots of pain in spots,
Especially near her bladder.
If only Jack would turn his back
She'd be a little gladder.

A Pulsatilla Poem

She had a try, but was too shy
To tell him of her worry;
But while she sat, it worsened. That
Then made her have to hurry.

The maiden soft then stood – and coughed!
Alas! Calamity!
Nasty wet knickers and cruel Jack's sniggers
Complete her misery.

The burning pain then came again,
Poor Jill began to cry.
Jack, with a shrug, gave her a hug
And the tears began to dry.

The Ballad of Jack and Jill

Jill would pout and be put out
Unless Jack went to get her
Something nice: a creamy ice
To make her feel better.

Jack said no, he wouldn't go,
So Jill began to whimper.
Jack, tho' cross, knew who was boss –
He'd rather have her simper.

Jill had her dream, a pink ice-cream,
She smiled from ear to ear,
Till came again a tummy pain
And with it: diarrhoea!

An added curse, to make life worse,
They had a chilly shower.
Nothing nearby could keep them dry,
They scampered to a tower.

A Pulsatilla Poem

Sore ears and eyes, complete with styes,
And poor Jill's other plight is
Catarrh like cream and pain, which means
Right-sided sinusitis.

Wet shoes which slip made poor Jack trip.
Oh, what a horrid shock!
Jack fell down and broke his crown
Upon a jagged rock.

Jill sat and moped. Could she have coped
If she had had some plaster?
With her fears came further tears
In facing this disaster.

Up Jack got and home did trot
As fast as he could caper.
He went to bed to mend his head
With Arnica and brown paper.

The Ballad of Jack and Jill

Pail and pack nigh broke Jill's back,
She slithered down the slope.
In her grief no tear relief,
She sat there without hope.

Then who should come but Jill's own Mum
And saw her looking blue,
Covered with mud, and stained with blood,
Pink mottled purple too.

She took Jill home to a bath of foam,
And sorted out her muddle;
In an hour or two as good as new
With PULSATILLA and a cuddle.

A Pulsatilla Poem

Tho' Jill did fail to fill the pail
Of water for her mother,
Another day, another way
To help her, she'd discover.

Jack, on the mend, has found a friend
Who isn't half as zealous.
They're off to play football all day,
Which leaves Jill feeling jealous.

Jack and Jill went up the hill
To fetch a pail of water.
Jack fell down and broke his crown,
And Jill came tumbling after.

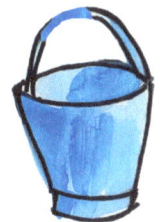

The Confessions of Clara Cuttlefish

A Sepia Poem

I couldn't care less
My life's a mess
There's too much stress
And strain.

Work in my prime
Made life sublime
Promotion time-
Fast lane!

My figure sleek
And freckled cheek
With dress sense chic
Would earn

The Confessions of Clara Cuttlefish

A yearning glance.
I could entrance
With every glance
And turn.

In routine mired
I soon grew tired
And so was fired -
Me, sacked!

Despite my dread
Agreed to wed -
My independence shed,
A fact!

The home then mine
With skilled design
And objects fine
I filled,

But joy refined
Was undermined
By routine grind,
And still'd.

A Sepia Poem

Despite our caution
Came an abortion.
Wretched our portion!
Reviled!

Each pregnancy
A penalty
Of ills. To be
With child.

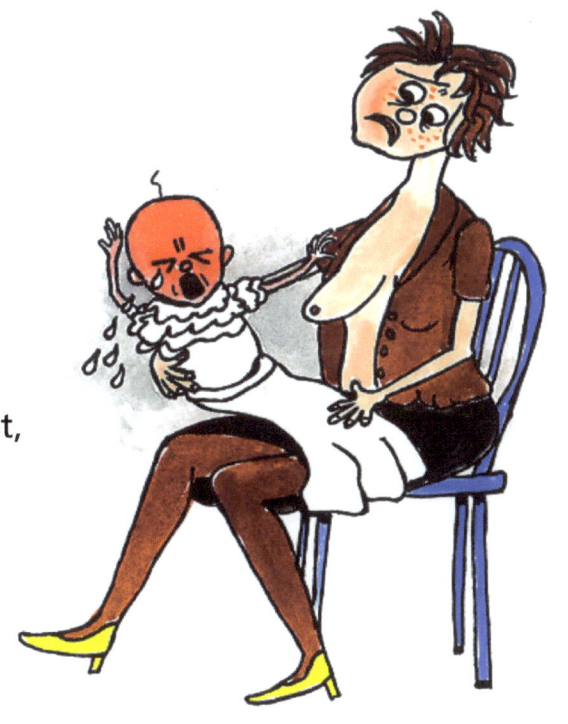

I failed the test:
They spurned my breast,
I was depressed,
I cried.

Life takes its course,
Each child's a source
Or rage, remorse -
I tried

To hide with mask
How huge each task
Which they would ask
Could be.

One other hell
Was there as well,
Perpetual smell
Of food.

I was quite torn asunder
By woman's monthly blunder
(Made worse, I thought by thunder)
So I'd brood.

Life without rhyme,
Tired all the time,
It seemed a crime
To me.

At first I fought.
The more I sought
Husband's support
I paid:

His sole life tape
Perpetual rape
Could I escape?
I prayed ...

A Sepia Poem

Relief? Long walk
And brisk; no talk,
No touch. Uncork
Cold drink.

By friendships ended
I felt offended,
Tho' not intended,
I think.

If I'm consoled
I cringe, go cold;
I won't be told,
Cause strife.

I came to see
I'd rather be
Anything but
A wife.

So in due course
Came my divorce.
Alas! no source
Of joy.

My need, all else above:
To be loved and to love -
Who would this worn-out glove
Employ?

The Confessions of Clara Cuttlefish

The mirror shows
My saddle nose,
Brown-blotched it grows
Like mould.

Sallow skin, thin hair,
Narrow hips, breast spare,
Droop'd pose, paunch rare,
Look old.

By body betrayed,
My ills arrayed,
Macabre parade
I'll paint:

Left in the lurch
By God, in church
I kneel and search -
And faint!

Most wretched cause!
By nature's laws
With menopause
I'm cursed.

Unwelcome blush
From flaming flush,
Sour sweats which gush,
And thirst.

A Sepia Poem

I'm flopping, flagging,
Eyelids sagging,
Pelvis dragging
Down

Upward stabs,
Throbbing jabs
My vulva grabs,
And crown.

Rotten to the core,
Foul discharge made me sore,
Yellow, brown and thin, and more
I've got

Piles third degree, perhaps is
More like fore and aft prolapses.
Weight of ball to fill the gaps is
My lot.

Now I'm alone
In fear. I moan.
Self-pitying crone
Am I.

In ocean I'll sink,
In dark crevice shrink,
In fog of brown ink
I'll escape, cause maximum confusion,
Catch my prey and get my own back!

The Rhyme of Ivy Rusto
A Rhus Toxicodendron Poem

A patient came, one Ivy Rusto,
Played her tennis with great gusto,
Very much enjoying it,
Even though she wasn't fit.

The wind grew cold, thick clouds rolled o'er
The court, yet Ivy called for more;
The thunder rolled, great drops of rain
Began to fall, but all in vain

Did Ivy's partner call out "Halt!"
The aftermath was Ivy's fault.
Upon the slippery court she fell:
"My ankle! Help!" did Ivy yell.

The Rhyme of Ivy Rusto

It hurt a lot and soon grew stiff
And swollen too. "If only - IF!
We'd stopped when Bryony had said
But now I'll have to take to bed."

Alas, that move brought yet more grief,
All through the night, without relief
No comfort lying there was found,
She tossed and turned and moved around

In restless, restless, RESTLESS state,
Her muscles all began to grate;
More pain and stiffness, in her back,
With every turn it seemed to crack.

Tho' every movement made her ill
She was much worse if she kept still.
Stretching she seemed to have to do,
Which helped, a bit – Catch 22!

Getting ill, combined with menses,
Exacerbated Ivy's herpes,
Blisters which, with every turn
Made her itch, and sting and burn.

For days poor Ivy suffered so,
Her friends would visit, come and go,
She wept and felt herself forsaken,
(Although in that she was mistaken).

A Rhus Toxicodendron Poem

Her fears at night, of evil, murder,
Torment brought. Misfortune further
Than she feared: The world's on fire
In fitful dreams. "It will require
Oceans and icebergs, without doubt
For weeks and months to put it out."

In other dreams the fields she'd roam,
And feel she was away from home
Tho' she was not. Poor Ivy wept
And fidgeted until she slept.

"I'm weary of this life," she'd cry,
"It's loathsome, and I'd rather die
Than live like cripples of this ilk.
I'm thirsty: please! Some nice cold milk."

And better still, there came a box
Of pills, whose label said: "Rhus-tox.
Please take, according to the letter."
She did - and soon was quite, quite better.

The Collapsed Man
A Carbo-Veg Poem

Call 999, the man's collapsed
I think that he is going to die.
His body's icy cold and blue
And wet with sweat, I wonder why?

He's cold, not only in his trunk,
His breath and tongue are icy too,
And he's confused. Perhaps he's drunk?
He drank too much at lunch, it's true.

It's worse than that. He's on the edge
Of life. Quick! Give him Carbo-veg.
Just watch him, breathless, cough and wheeze,
He craves fast fanning and a nice cool breeze,
And something else that should be noted:
His face and abdomen are bloated.

The Collapsed Man

A burning fire, a conflagration,
Somewhat eased by eructation,
Is sitting in the poor man's belly.
He's old, and can't help being smelly.

He's sore with diarrhoea and flatus
In huge amounts from his meatus,
Caused by the simplest foods. He pays
If he eats fat, or rich arrays.

In fact, he rapidly gets fractious
If he eats anything that's scrumptious.
You're feeding him? Then there's a trick:
The foods he craves for make him sick.

His diarrhoea and his wind
Are worse when hot sunshine is there,
But very strangely though, his mind
Deteriorates in warm wet air;
He's dull, more likely to forget,
Can't rouse himself, which he'll regret.

*"To teach us you have made a pledge.
What sort of man is Carbo-veg?"*
First of all I need to say
The source: the silver birch-en tree
On fire. The energy's burnt away,
There's little left, as you will see.

And so, in answer to your question,
His mind and body are both sluggish,
Totally low, not just digestion.
He is a sort of human rubbish.

The Collapsed Man

When young he once was very ill;
He never quite got over it.
That bad antibiotic pill
Has left him far from fully fit.

He's worse when going for a walk.
He's much too fat as you can see,
And half the time he cannot talk,
At least, not very easily.

Anxious, frightened, closing eyes
Are in his hot (and bloated) face;
In open air or sleep he cries.
He's often doubtful of his place.
His mind plays tricks on him, I fear,
He feels his head is tied up tight,
He'll phantoms see, or voices hear,
Finds strangers near a horrid blight.

A Carbo-Veg Poem

A businessman he tries to be,
And trying it may make him cross.
He cannot concentrate, you see,
So business may well end in loss.

Finding everything a struggle,
He's often living in a muddle.
Indifferent, he doesn't care,
He doesn't fight you for his share,
And if you press he'll soon give way -
You see, he doesn't care enough
To insist and have his say
If the going's getting tough.

"What happened to our 'collapsed man'?"
I give my honest pledge:
No. You should guess. I know you can.
"… He's cured by Carbo-veg?"
Yes! YES!

No Joy for Jasmine!
A Gelsemium Poem

Oh, dear! What can the matter be?
Poor young Jasmine's rushed to the lavatory.
She's been going from Monday to Saturday,
Day of her driving test.

"Come now, Jasmine, you can drive very well,"
Did Jasmine hear? Her instructor couldn't tell.
Shook like a leaf, as if she were going to hell,
Not just her driving test.

Poor wee Jasmine's full to the brim with fear,
Weak-kneed, falling, wishing she wasn't here,
Memory gone, neither thinking nor vision clear,
Couldn't respond to th'examiner's words of cheer,
Shrivelled and oh so dry, under his gaze and leer
Failed her driving test.

No Joy for Jasmine!

Oh, dear! What does the season bring?
Warm sunny days, soft sweet showers in spring,
Making things grow and the sleek blackbird sing
But not Jasmine, it gives her a chill.

It starts in her hands, or goes up and down her back.
Chill, heat, sweat, more chill, poor Jasmine just can't keep track,
Muscles all sore, and the outlook is turning black.
Jasmine feels very ill.

No thirst, tho' she's dry, and her face a dull glow,
Weak and exhausted, her pulse soft and slow,
She wants you to hold her because she shakes so,
That's Jasmine with a chill.

Matchsticks she needs to prop open her eyes,
Pupils dilated, as if in surprise.
"Please turn the light off, it hurts me!" then cries
Jasmine in her chill.

A Gelsemium Poem

First there's coryza affecting her nose,
The feeling that from it some hot water flows,
The inside's all swollen, it's blocked when she blows.
That's Jasmine with a cold.

Coryza's not all, pharyngitis will follow,
Rough, burning, sore, and a lump she can't swallow,
Tongue heavy and trembling, and thick-coated yellow,
With difficult speech as she's propped on her pillow,
Jasmine's not feeling bold.

Oh, dear! We fear Jasmine's going west,
Voice nearly gone, we find when we make a test,
Cough tickly, sore and dry, and a lump in her chest,
Bronchitis now, her virus a real pest,
Jasmine seems to fold.

No Joy for Jasmine!

Oh, dear! Here comes the sun,
Jasmine sunbathing, more trouble's begun,

She's suddenly blind, so it's panic, not fun,
Returns, but with headache, an occipital one.
Jasmine, off to your bed!

It's bursting, it's swollen, it's like a tight band,
She must rest with head raised, in the dark, and not fanned.
Wrapped warm; no thinking, the air must be canned
To soothe poor Jasmine's head.

Her headache creeps forward and finds her eyeballs,
Still swollen, congested, her eyelid still falls,
Still dull red, besotted, appearance appals,
As Jasmine lies in her bed.

Whatever her illness, we always can see
When she's due to get better, she goes off to pee,
Not once, and not little, but repeatedly!
And this too is strange, now, don't you think?

She's depressed, suicidal, but doesn't complain
Of all her afflictions, the pains that remain
And the trembling and weakness will all come again
With bad news, excitement, or even in rain.
Making Jasmine unwell.

Despite being dry-mouthed, she doesn't want drink.
And with copious perspiring she'll very soon be
As healthy as you, and as healthy as me.

For more of Dr John English's creative learning materials for homeopaths, including quizzes, mind maps, poems and dramatic presentations, plus a regular blog see:

www.EnjoyLearningHomeopathy.co.uk

www.ingramcontent.com/pod-product-compliance
Lightning Source LLC
Chambersburg PA
CBHW041100180526
45172CB00001B/46